Extreme Weather

# THUNDERSTORMS

Martha London

DiscoverRoo
An Imprint of Pop!
popbooksonline.com

abdobooks.com

Published by Pop!, a division of ABDO, PO Box 398166, Minneapolis, Minnesota 55439. 

Printed in China.

052019
092019

THIS BOOK CONTAINS RECYCLED MATERIALS

Cover Photo: Shutterstock Images

Interior Photos: Shutterstock Images, 1, 5, 6, 11, 18, 19, 25, 28, 29 (bottom); iStockphoto, 7, 8–9, 10, 13 (lightning), 15, 16, 17, 22, 24, 29 (clouds), 29 (checkmarks), 29 (car), 29 (house), 29 (trees), 29 (flooded house), 29 (plug), 29 (microwave), 29 (television), 30; Claus Lunau/Science Source, 12–13 (storm cloud); Mike Olbinski/Alamy, 21; LLC/Zuma Press/Newscom, 23; NOAA, 27, 31

Editor: Connor Stratton

Series Designer: Sophie Geister-Jones

**Library of Congress Control Number: 2018965059**

**Publisher's Cataloging-in-Publication Data**

Names: London, Martha, author.

Title: Thunderstorms / by Martha London.

Description: Minneapolis, Minnesota : Pop!, 2020 | Series: Extreme weather | Includes online resources and index.

Identifiers: ISBN 9781532163951 (lib. bdg.) | ISBN 9781532165399 (ebook)

Subjects: LCSH: Thunderstorms--Juvenile literature. | Electrical storms--Juvenile literature. | Weather--Juvenile literature. | Rainstorms--Juvenile literature.

Classification: DDC 551.55--dc23

# WELCOME TO DiscoverRoo!

Pop open this book and you'll find QR codes loaded with information, so you can learn even more!

Scan this code* and others like it while you read, or visit the website below to make this book pop!

**popbooksonline.com/thunderstorms**

*Scanning QR codes requires a web-enabled smart device with a QR code reader app and a camera.

# TABLE OF CONTENTS

## CHAPTER 1

# HOW THUNDERSTORMS FORM

A thunderstorm happens when clouds make rain and **lightning**. A thunderstorm needs warm, moist air to form. This warm air rises and meets cold air above. The warm air cools.

WATCH A VIDEO HERE!

*Thunderstorm clouds can be as high as the tallest mountain on Earth.*

Then it **condenses** into drops of water. These drops stick to tiny bits of dust in the air. Together, they form a cloud.

Water makes the bits of dust heavy. If they are too heavy, they fall out of the cloud. These falling drops are rain. The falling rain meets rising warm air. When they meet, an **electric charge** builds. If clouds have enough electric charge, lightning can form.

**DID YOU KNOW?**

**The heat from lightning makes the air around it grow and shrink quickly. This change creates the sound of thunder.**

Some thunderstorms have lightning but no rain. These storms are called dry thunderstorms.

A thunderstorm forms from rising warm air and falling cold air. These two air systems have different temperatures.

*Lightning can sometimes be seen up to 100 miles away.*

The larger the difference in these two temperatures, the stronger a thunderstorm can become.

Sometimes a storm's rising air also has strong, wild winds. These rising winds can make the storm cloud spin. This swirling storm is called a supercell thunderstorm. Supercells are the strongest type of thunderstorm.

*Many supercells look like anvils.*

The average thunderstorm lasts 30 minutes. But supercells can last for many hours.

# HOW STORM CLOUDS FORM

Clouds form when falling cold air meets rising warm air. The types of clouds that cause thunderstorms are often very tall. These clouds are sometimes called thunderheads.

## CHAPTER 2

# WHEN AND WHERE

Thunderstorms can happen at any time. But they are most common in the spring and summer. And they often happen in the afternoon. The ground is warmest

**DID YOU KNOW?**

**Approximately 1,800 storms are happening on Earth at any given time.**

*On average, thunderstorm clouds travel 30 to 40 miles per hour.*

LEARN MORE HERE!

at that time. This heat creates the rising warm air that storms need.

Thunderstorms can happen anywhere on Earth. But they are most common close to the **equator**. Temperatures there are hotter. Storms are also most common in **tropical** areas. The air there is usually moist.

Parts of Florida are tropical. This state receives more thunderstorms a year than most other states.

For these reasons, most thunderstorms in the United States happen in southern states. Storms are also common in central Africa and

*A thunderstorm hits Orlando, Florida.*

*Thunderstorms hit Java island in Southeast Asia almost every day of the year.*

South America. They often hit island nations in eastern Asia too.

**DID YOU KNOW?**

**Cimarron, New Mexico, has up to 110 thunderstorms every year!**

## CHAPTER 3

# DAMAGE

Thunderstorms can cause damage in many ways. For example, **lightning** can be deadly. It can make the air hotter than the sun. Lightning can cause trees to explode. It can even start wildfires.

LEARN MORE HERE!

*Lightning that starts a wildfire is often from a dry thunderstorm.*

**DID YOU KNOW?**

**There are approximately 100 lightning strikes on Earth every second.**

Thunderstorms can also bring harmful rain and wind. Heavy rains can flood roads and buildings. Some thunderstorms even make **hail**. Hail can hit cars, homes, and people.

*Hail can be smaller than a pea or as big as a softball.*

Storm winds can gust faster than 100 miles per hour. They can rip trees from the ground.

## LIGHTNING STRIKES!

In 2012, a thunderstorm in California broke records. The huge storm lasted more than four hours. Approximately 750 lightning strikes hit one area. Lightning struck buildings, the Golden Gate Bridge, and even an airplane.

Supercell thunderstorms bring stronger winds and rain than other thunderstorms. They can also last for several hours longer. Supercells can even make **tornadoes**. Tornadoes have high winds that can destroy houses and other buildings.

*In 2019, powerful tornadoes hit a town in Alabama. One tornado's winds traveled 170 miles per hour.*

CHAPTER 4

# STAYING SAFE

Scientists study weather patterns to tell when a thunderstorm might come. If a storm with strong winds or **hail** is possible, they send out a thunderstorm watch. If a strong storm has started, they send a thunderstorm warning.

COMPLETE AN ACTIVITY HERE!

*Satellites in space take pictures of Earth. Scientists use these pictures to study the weather.*

*Being under a tree increases a person's chances of being hurt by lightning.*

During a thunderstorm, people move inside if they can. If they are outside, people avoid hiding under trees. Tall objects attract **lightning**. People also stay out of flooded areas. And they do not take baths or showers. Water can shock people if lightning strikes it.

## Thunderstorm Safety Checklist

- ✓ Find shelter inside a building or car.
- ✓ Do not hide under tall objects.
- ✓ Avoid flooded areas.
- ✓ Unplug appliances.
- ✓ Follow the news for weather updates.

# MAKING CONNECTIONS

## TEXT-TO-SELF

Are thunderstorms common where you live? Why or why not?

## TEXT-TO-TEXT

Have you read books about other kinds of storms? How are they similar to thunderstorms? How are they different?

## TEXT-TO-WORLD

Lightning can be a dangerous part of a thunderstorm. Why is it safer to stay away from tall objects, such as trees?

# GLOSSARY

**condense** – to change from a gas to a liquid.

**electric charge** – the amount of energy in something.

**equator** – an imaginary line that runs around the middle of Earth.

**hail** – frozen rain that forms in a thunderstorm.

**lightning** – an intense release of energy in the air that causes a flash of light.

**tornado** – a violent storm whose winds spin around on land.

**tropical** – having to do with a place where the weather is usually warm and wet.

# INDEX



## ONLINE RESOURCES

**popbooksonline.com**

Scan this code* and others like it while you read, or visit the website below to make this book pop!

**popbooksonline.com/thunderstorms**

*Scanning QR codes requires a web-enabled smart device with a QR code reader app and a camera.